20-MINUTE (OR LESS) MEME HACKS

SHEELA PREUITT

Lerner Publications ◆ Minneapolis

To Dr. V.S. Viswanath, for his brilliance and generosity.

Online Safety

Children should always ask permission before going online, especially when using a new website. Stay safe online by protecting your identity. Never share personal information such as your full name, address, or email address online. Think carefully about posting personal photos of yourself, and don't post pictures of friends or family without permission.

Lerner Publications Company
An imprint of Lerner Publishing Group, Inc.
241 First Avenue North
Minneapolis, MN 55401 USA

For reading levels and more information, look up this title at www.lernerbooks.com

Main body text set in Conduit ITC Std Regular 13/17
Typeface provided by ITC

Library of Congress Cataloging-in-Publication Data
Names: Preuitt, Sheela, author.
Title: 20-minute (or less) meme hacks / Sheela Preuitt.
Other titles: Twenty minute (or less) meme hacks
Description: Minneapolis : Lerner Publications, 2020. | Series: Vidcode coding hacks | Includes index. | Audience: Ages 8–12. | Audience: Grades 4–6. | Summary: "Using the Vidcode platform and just a few lines of code, readers can combine pictures and text to create hilarious memes! The projects are customizable and build a foundation of coding skills"— Provided by publisher.
Identifiers: LCCN 2019052414 (print) | LCCN 2019052415 (ebook) | ISBN 9781541597198 (library binding) | ISBN 9781728401447 (ebook)
Subjects: LCSH: Computer graphics—Programming—Juvenile literature. | Memes—Juvenile literature.
Classification: LCC T385 .P746 2020 (print) | LCC T385 (ebook) | DDC 006.6—dc23

LC record available at https://lccn.loc.gov/2019052414
LC ebook record available at https://lccn.loc.gov/2019052415

Manufactured in the United States of America
1 – CG – 7/15/20

TABLE OF CONTENTS

CODE YOUR MEME

Have you ever wondered how to create hilarious memes? You can do this in 20 minutes or less by **coding**! Using the Vidcode platform, combine text and graphics with just a few lines of code. You'll be **programming** before you know it!

Coding takes practice. As you learn, you might give the computer a wrong instruction and end up with something you don't like. When this happens, don't worry! Stop, think, and change your code to get the result you want.

Ready to show off your creativity?

To complete each project in this book, you can work online at the Vidcode Lerner sandbox. You will be able to save and customize your personal projects.

vidcode

GO HERE TO GET STARTED!

qrs.lernerbooks.com/Memes

MAKE YOUR OWN MEME

PROJECT TIME: 15 MINUTES

JavaScript is a **programming language**. Use it to combine pictures and text and create a meme!

CODING SKILLS

- objects
- properties

PROJECTS:
qrs.lernerbooks.com/Memes

Click the Back and Next buttons in the Instructions panel at any time to unlock effects and access useful tips and tricks.

1. Click on an image from the Backgrounds tab. The image will appear on the **canvas** and as two lines of code in the **code editor**.

2. Click on an image from the Graphics tab. The graphic will appear on the canvas and as one line of code in the code editor.

3. Make your graphic bigger or smaller by using the 'scale' **property**. Make the graphic move left, right, up, and down by using the 'x' and 'y' properties. Type the code below into your code editor to begin experimenting:

```
my_graphic1.scale = 1.5;
my_graphic1.x = 50;
my_graphic1.y = 250;
```

Change the property **values** to make your graphic exactly how you'd like.

1

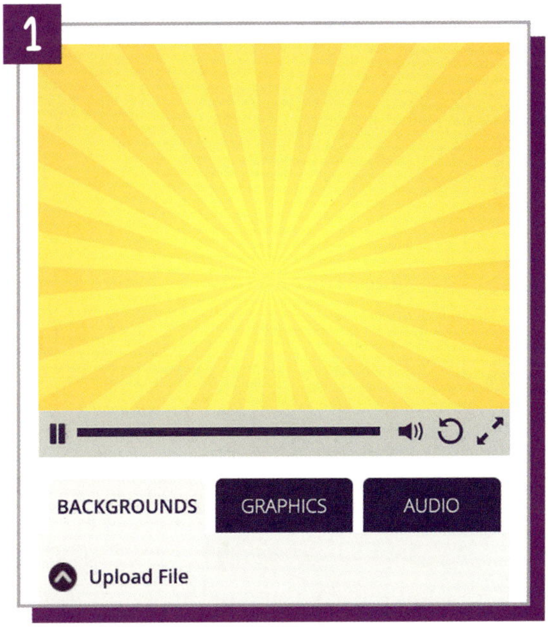

2

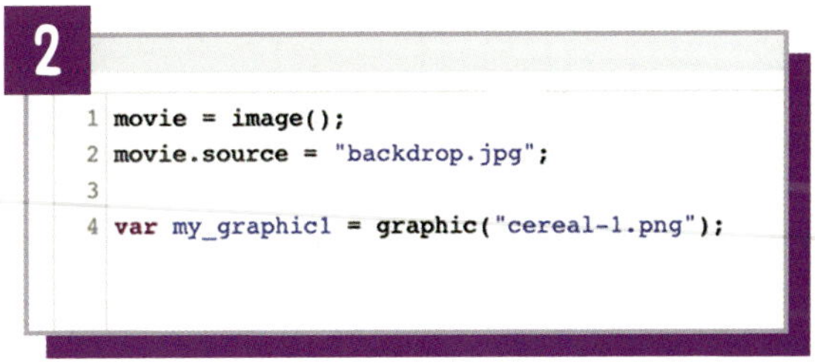

4. Follow the prompts and answer questions in the Instructions panel until a text block appears in the Effects tab. Drag this block into the code editor.

5. The code editor should show the line:
`var my_text1 = text("hello");`
Select the word "hello" to change it to any word or words you like. Do not delete the quotation marks.

6. Add properties to your text to change its size, color, or position. Type in the code below to begin experimenting:
```
my_text1.size = 100;
my_text1.color = "red";
my_text1.x = 50;
my_text1.y = 10;
```
Change the values of these properties until your text is as you want it!

5

```
movie = image();
movie.source = "backdrop.jpg";

var my_graphic1 = graphic("cereal-1.png");

my_graphic1.scale = 1.5;
my_graphic1.x = 50;
my_graphic1.y = 250;

var my_text1 = text("hello");

```

6

```
movie = image();
movie.source = "backdrop.jpg";

var my_graphic1 = graphic("cereal-1.png");

my_graphic1.scale = 1.5;
my_graphic1.x = 50;
my_graphic1.y = 250;

var my_text1 = text("I'm so hungry!");

my_text1.size = 100;
my_text1.color = "red";
my_text1.x = 50;
my_text1.y = 10;

```

CODING TIP!

In JavaScript, properties describe or belong to an **object**. Text and images are objects. So, properties alter how text and images look and behave.

DOGE MEME

PROJECT TIME:
15 MINUTES

Variables are containers. They can hold objects, like text and graphics. Use variables to code your very own **doge** meme!

CODING SKILLS

- variables
- objects
- properties

PROJECTS:

qrs.lernerbooks.com/Memes

1. Drag a text block from the Effects tab into the code editor. The text block will appear as one line of code. In the code, `var` stands for variable, `my_text1` is the variable name, and `text` is the object.

2. Select the variable name and change it to `dogeText`.

3. Apply properties to the text by referencing the variable name. Change the color, size, and message of your text by adding and adjusting properties.

4. Move your text left, right, up, and down by adding 'x' and 'y' properties.

5. Click on the doge-1 image from the Graphics tab. The graphic will appear as one line of code in the code editor.

6. Select the variable name and change it to `dogeImage`. Add 'x' and 'y' properties to your graphic to change its position. Use the scale property to change its size.

3

```
var dogeText = text("hello");
dogeText.message = "So much fun coding!";
dogeText.color = "yellow";
dogeText.size = 50;
```

4

```
var dogeText = text("hello");
dogeText.message = "So much fun coding!";
dogeText.color = "yellow";
dogeText.size = 50;
dogeText.x = 50;
dogeText.y = 50;
```

6

```
var dogeMessage = text("hello");
dogeMessage.message = "So much fun coding!";
dogeMessage.color = "yellow";
dogeMessage.size = 60;
dogeMessage.x = 50;
dogeMessage.y = 50;
var dogeImage = graphic("doge-1");
dogeImage.x = 50;
dogeImage.y = 100;
dogeImage.scale = 1.2;
```

BESTIE GREETING CARD

Create a cool video greeting card for your best friend using graphics, text, and **filters**.

PROJECT TIME: 20 MINUTES

CODING SKILLS

- variables

PROJECTS:

qrs.lernerbooks.com/Memes

1. Click on a video from the Background tab.

2. From the Effects tab, drag and drop some filters into the code editor.

3. Below the lines of code in your code editor, type:
   `movie.playbackRate = 0.5;`
   Change the value in the code above to make the video play faster or slower.

4. From the Graphics tab, click to add a graphic on top of the video. The name of the variable holding your new graphic will be `my_graphic1`.

2

```
movie = video();
movie.source = "crafts.mp4";

tint("red", 20);
kaleidoscope();

```

3

```
movie = video();
movie.source = "crafts.mp4";

tint("red", 20);
kaleidoscope();

movie.playbackRate = 0.5;
```

4

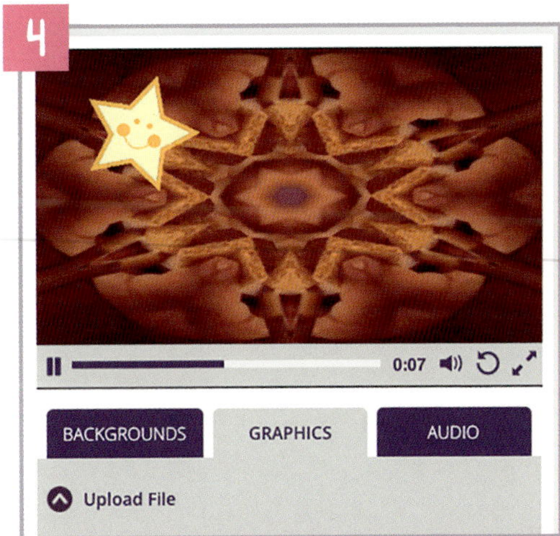

5. Make the graphic bigger or smaller by adding the scale property. Move the graphic left, right, up, and down by applying 'x' and 'y' properties:

```
my_graphic1.scale = 1.5;
my_graphic1.x = 300;
my_graphic1.y = 50;
```

6. Answer quizzes in the Instructions panel to unlock the text block. Then drag the text block from the Effects tab to the code editor.

7. Add 'x', 'y', color, size, and message properties to the text to say something special!

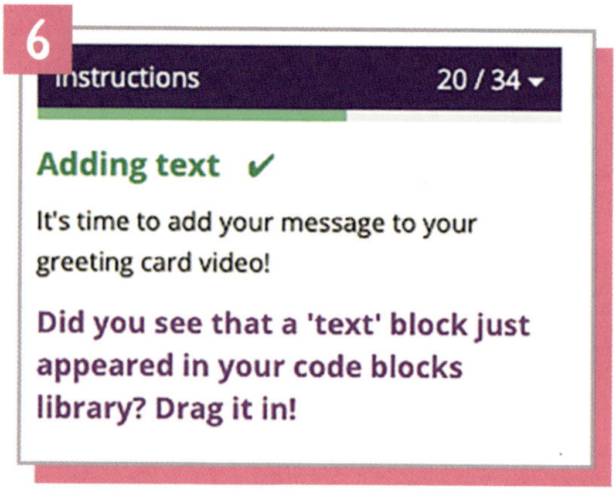

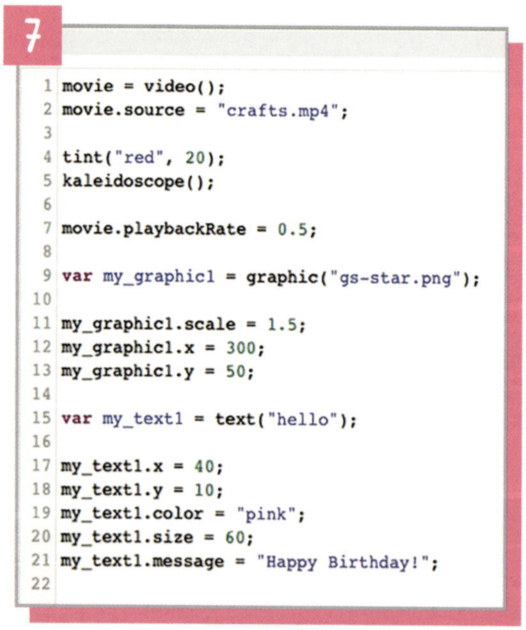

CODING TIP!

Variables are key to writing programs. Variables make it fast and easy to add and change code. This saves programmers precious time and mental energy!

DESIGN A TYPEFACE

Most of the letters of the English alphabet can be drawn using circles and lines. Code lines and circles to create your own **typeface**!

PROJECT TIME:
20 MINUTES

CODING SKILLS

- variables
- objects
- properties

PROJECTS:
qrs.lernerbooks.com/Memes

1. From the Effects tab, drag and drop a line block into the code editor. You will see the line of code:
   `var my_line =`
   `line(0,0,100,100);`
   In this code, `var` stands for variable. The name of the variable is `my_line` and the object the variable holds is the `line`.

1

2. Every line has a start and end. The position of these points on your canvas is defined by the numbers shown in parentheses in your code. The first two numbers, 0 and 0, represent the 'x' and 'y' position of the start point. The next two, 100 and 100, represent the 'x' and 'y' position of the end point. Adjust these values to move and adjust lines!

2

```
var my_line = line(100, 100, 200, 300);
```

3. The letter *V* is made from two lines. Play around with the values in parentheses to create the first line of this letter.

CODING TIP!

In code, you must always type a variable name the same way. `my_line` is different from `My_line` or `myline`.

4. Drag a second line block into the code editor. Adjust the values in this line of code to complete the *V*.

5. Select the two variable names. Change the first variable to `my_line1` and the second to `my_line2`.

6. Change the color of a line by adding a fifth **argument** to the list in parentheses:

```
var my_line1 = line(100,
100, 200, 300, "white");
```

7. Change the thickness of a line by adding a sixth argument to the list in parentheses:

```
var my_line1 = line(100,
100, 200, 300, "white", 5);
```

Make this value bigger or smaller to make the line thicker or thinner.

8. Drag a circle block into the code editor from the Effects tab.

```
var my_circle = circle(50,
50, 100);
```

In this code, the name of the variable is `my_circle`. The object the variable holds is the `circle`.

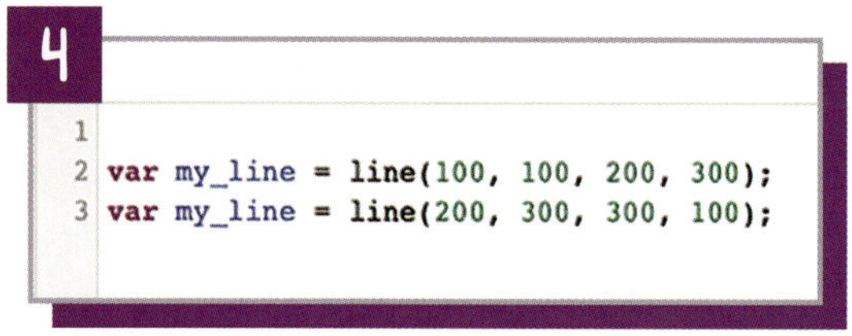

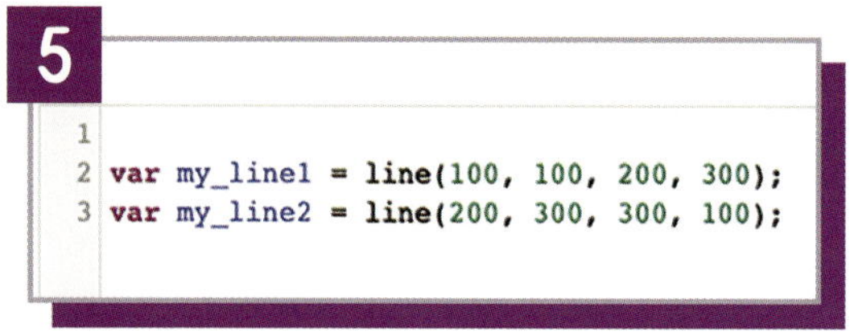

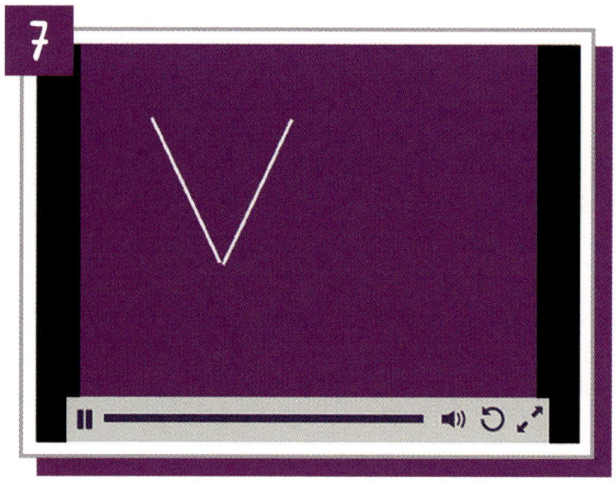

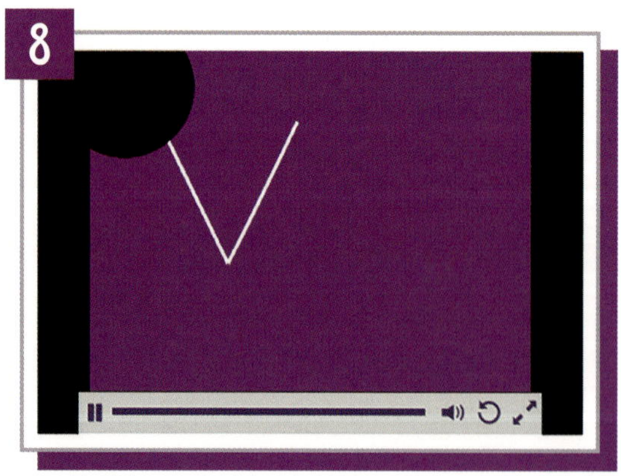

9. The first two numbers in parentheses, 50 and 50, represent the 'x' and 'y' position of the circle's center. The number 100 represents the **radius** of the circle. Alter these values to change the circle's size and location.

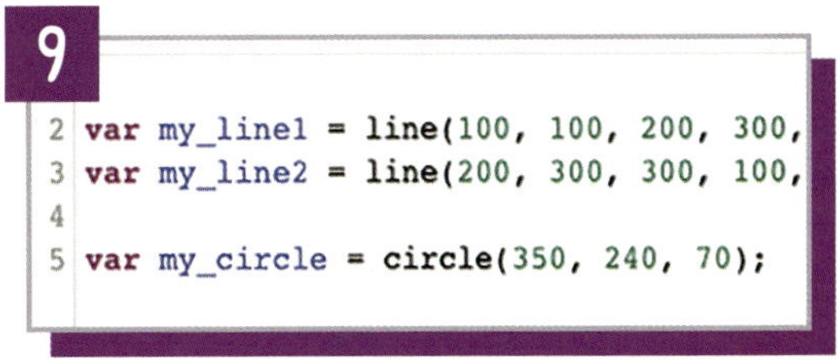

10. Add values inside the parentheses to alter the color, border color, and border width of the circle:

```
var my_circle = circle(150,
150, 50, "black", "white",
10);
```

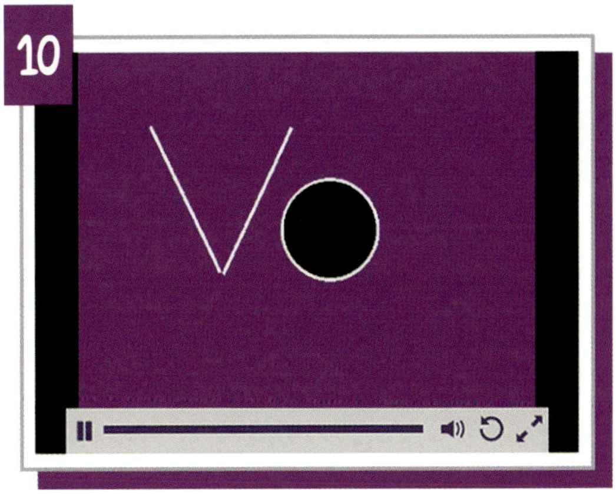

11. Make the letters *P*, *B*, *R*, and others by combining lines with circles. Create as many letters as you can in your awesome new typeface!

HACK IT!

The order of arguments inside parentheses is important. Type the code `line(0, 0, 100, 100, "red")` in the code editor. Then move "red" from last place to first inside the parentheses. What happens?

BEHIND THE CODE

It may not be obvious at first what your code is instructing. But every line of code has a purpose!

The letter "x" refers to the x axis, which is the width of the canvas.

The number "640" represents the number of pixels that make up the x axis.

The letter "y" refers to the y axis, which is the height of the canvas.

The number "360" represents the number of pixels that make up the y axis.

```
var my_line1 = line(100, 100, 200, 100, "lightblue", 7);
var my_line2 = line(100, 100, 100, 300, "lightblue", 7);
var my_line3 = line(100, 200, 200, 200, "lightblue", 7);
```

The fifth argument in a line **function** is color.

```
var my_circle = circle(300, 230, 70, "red", "lightblue", 15);
```

The fourth argument in a circle function is fill color.

```
var my_line4 = line(400, 165, 530, 300, "lightblue", 7);
var my_line5 = line(530, 165, 400, 300, "lightblue", 20);
```

The sixth argument in a line function is thickness.

WIGGLEGRAM

Wigglegrams are images that move back and forth to make a 3-D effect. Make a Wigglegram by storing image files in an **array**.

PROJECT TIME:
10 MINUTES

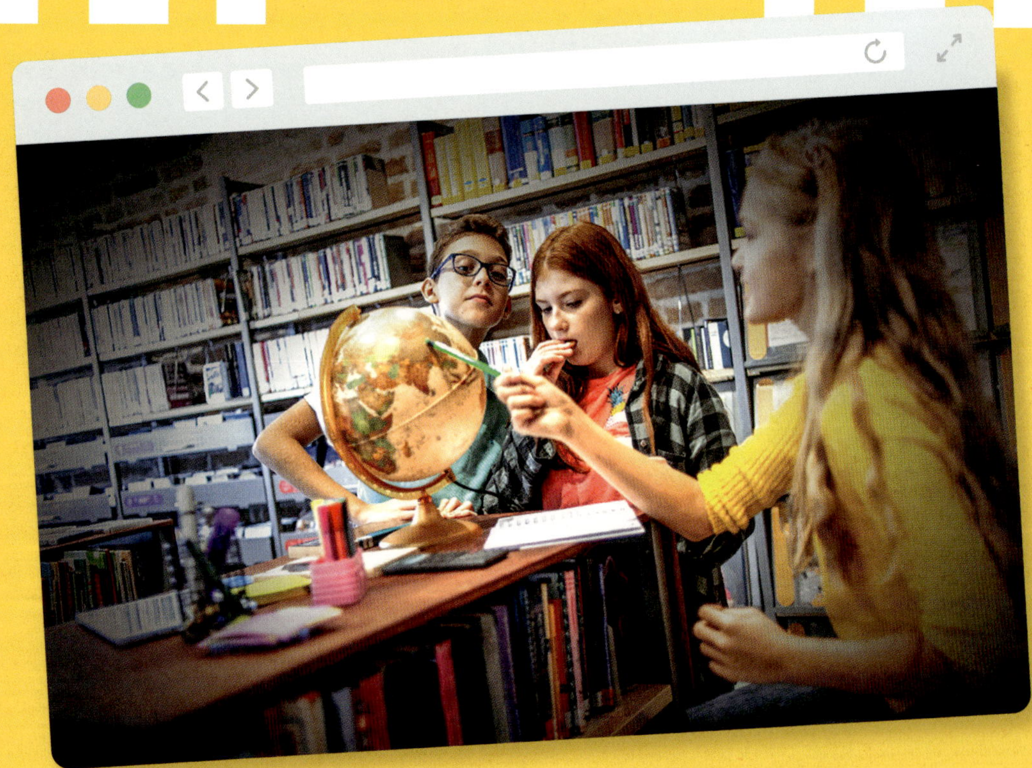

CODING SKILLS

- arrays

PROJECTS:

qrs.lernerbooks.com/Memes

1. Take two photos of your classroom and save them to your computer or tablet.

2. Click on Upload File in the Frames tab. Find and select the two photos of your classroom, so they appear as **thumbnail** images.

3. Click on the two images you uploaded. You should see a line of code beginning with:
`movie.frames`
This code shows your photo file names listed in brackets.

4. Under this, type:
`movie.interval = 500;`
This will make your video alternate between the two photos.

5. Make the value smaller or larger to change how quickly the **frames** alternate.

6. Change the look of your Wigglegram by dragging in filters from the Effects tab.

3

```
movie = stopmotion();

movie.frames = ["my_class1.jpg",
                "my_class2.jpg"];

```

4

```
movie = stopmotion();

movie.frames = ["my_class1.jpg",
                "my_class2.jpg"];

movie.interval = 500;
```

6

HACK IT!

A Wigglegram's movie frames are stored as an array. An array is like a list written inside brackets and separated by commas.

VIDEO DJ

PROJECT TIME: 20 MINUTES

Create a music video that changes tracks as you tap your keyboard!

Pressing a key on the keyboard or clicking a button on a web page are "events." JavaScript can be coded to respond to events in a specific way. In this project, when any key is pressed, JavaScript changes the audio file.

CODING SKILLS

- audio
- keyboard events

PROJECTS:

qrs.lernerbooks.com/Memes

1. Click on a video from the Background tab.

2. Click on any music file except "retro.mp3" from the Audio tab. The audio file will appear as one line of code in the code editor. The music will play automatically with your video.

3. On the same line as the audio file name, type `var my_audio =` This will create a variable to hold your audio file. The name of the variable is `my_audio`.

4. From the Effects tab, drag and drop the whenKeyDown **event listener** to your code.

5. Inside the whenKeyDown event listener, type the code: `my_audio.source = "retro.mp3";` Click on the video canvas. The code above will make the song change to `retro.mp3` when you press down on any key!

1

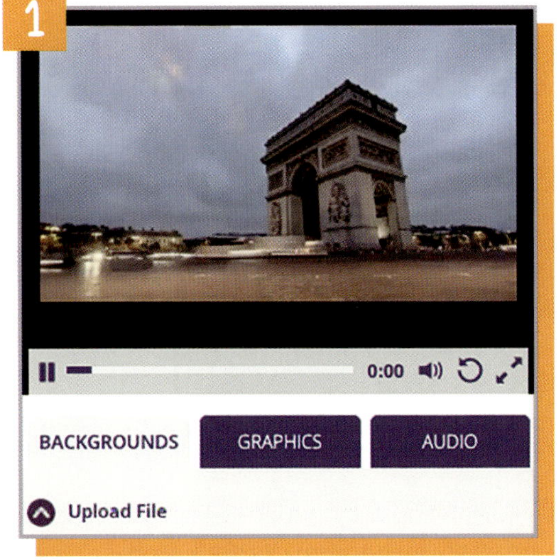

3

```
movie = video();
movie.source = "city_sunset.mp4";

var my_audio = audio("dance.mp3");
```

5

```
movie = video();
movie.source = "city_sunset.mp4";

var my_audio = audio("dance.mp3");

movie.whenKeyDown = function(key) {
  my_audio.source = "retro.mp3";
};
```

6. Above the whenKeyDown event listener block, create a new variable in the code editor called:
`var my_songs = [];`

7. Inside the square brackets of the `my_songs` variable, type more audio file names from the Audio tab. Separate the music file names by commas and enclose each file name in quotation marks. You've made an array. An array can hold a list of tunes!

8. Below the array, create a new variable called index:
`var index = [0];`

6

```
movie = video();
movie.source = "city_sunset.mp4";

var my_audio = audio("dance.mp3");
var my_songs = [];

movie.whenKeyDown = function(key) {
  my_audio.source = "retro.mp3";
};
```

7

```
movie = video();
movie.source = "city_sunset.mp4";

var my_audio = audio("dance.mp3");
var my_songs = ["dance.mp3", "electronic.mp3",
"retro.mp3", "rock.mp3", "funk.mp3"];

movie.whenKeyDown = function(key) {
 my_audio.source = "retro.mp3";
};
```

8

```
movie = video();
movie.source = "city_sunset.mp4";

var my_audio = audio("dance.mp3");
var my_songs = ["dance.mp3", "electronic.mp3",
"retro.mp3", "rock.mp3", "funk.mp3"];
var index = [0];

movie.whenKeyDown = function(key) {
  my_audio.source = "retro.mp3";
};
```

9. Change the audio source inside the whenKeyDown event listener to the name of your new array of songs, followed by `[index]`:
`my_audio.source = my_songs[index];`

10. On a new line inside the whenKeyDown event listener, add:
`index++;`
This will make the list of audio files in your array play in order one time.

11. To have the array play more than once, change `[index]` from step 9 to `[index%my_songs.length]`
Now you can DJ the night away with a never-ending playlist!

9

```
movie = video();
movie.source = "city_sunset.mp4";

var my_audio = audio("dance.mp3");
var my_songs = ["dance.mp3", "electronic.mp3",
"retro.mp3", "rock.mp3", "funk.mp3"];
var index = [0];

movie.whenKeyDown = function(key) {
  my_audio.source = my_songs[index];
};
```

11

```
movie = video();
movie.source = "city sunset.mp4";

var my_audio = audio("dance.mp3");
var my_songs = ["dance.mp3", "electronic.mp3",
"retro.mp3", "rock.mp3", "funk.mp3"];
var index = [0];

movie.whenKeyDown = function(key) {
  my_audio.source =
      my_songs[index%my_songs.length];
  index++;
};
```

HACK IT!

The punctuation // indicates the start of a note. Notes are ignored by JavaScript. So, anything you write after these symbols are just reminders to yourself of important details while coding.

SANDBOX FOR FREE CODING

Test your newfound skills and let your imagination guide your next incredible meme!

CODING SKILLS

- imagination
- exploration

PROJECTS:
qrs.lernerbooks.com/Memes

1. Click on a video clip from the Backgrounds tab.

2. Mix, match, and experiment with cool text and drawing options from the Effects tab.

3. Read through the Reference tab to see the different ways you can create cool coding projects with Vidcode. Cut and paste suggested snippets of code to see how it affects your work.

4. Click the "Save" button at the bottom of the code editor to save your project and come back to it later.

5. Click "Publish" to put your new project online for others to admire!

2

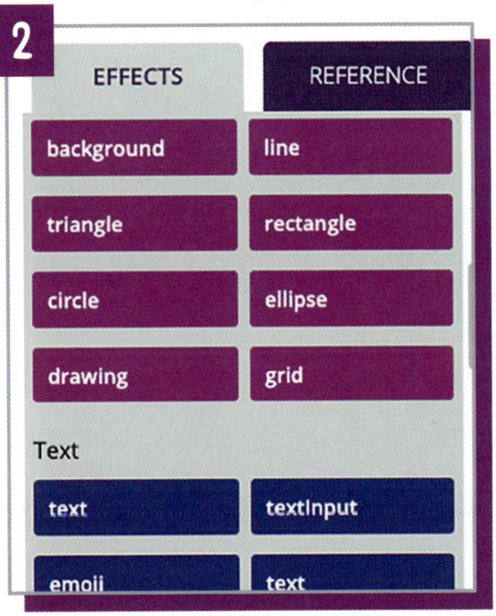

3

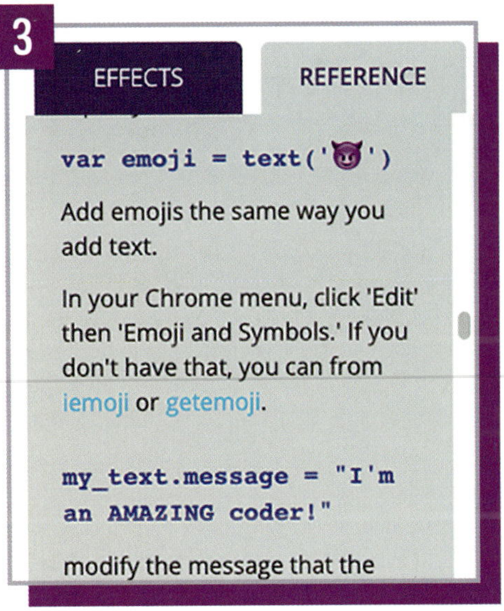

CODING TIP!

Imagination is essential to becoming an expert coder. Have fun coding and creating awesome projects to share with friends and family!

CONCLUSION

You are a coder who can now speak JavaScript! Take your skills to the next level by working on the Vidcode platform to continue learning by doing.

Have fun making awesome coding projects!

MORE ABOUT VIDCODE

Vidcode is a coding platform and curriculum that teaches computer science, object oriented programming, web programming, design, & JavaScript. The magic behind Vidcode has an academic term—computational media! This means learners can upload photos, illustrations, videos and audio—and manipulate them with code right away, allowing tweens and teens to connect computer programming to the media they interact with every day. Vidcode was founded by three women—an engineer, an educator, and an artist. We're now a global team that supports schools, libraries and individual learners in leveraging our 300+ hours of online coding tutorials.

Visit www.vidcode.com for more information.

GLOSSARY

argument: a specific value that acts as input to a function

array: a list of values

canvas: the top-right screen in Vidcode where your coding is displayed as pictures

code editor: the white box where your code is written

coding: writing instructions for a computer. Computer programs are coded.

doge: a type of meme that uses the image of the dog breed Shiba Inu

event listener: a program or function that waits for an event to occur. A mouse click or a key press is an event.

filter: a computer instruction that layers effects over an image or video

frame: an individual image in a sequence of images from a video or animation

function: a set of instructions to do a specific task

object: a container or an entity

program: a specific set of ordered commands for a computer to perform. To program is to write these commands.

programming language: a set of words and rules for giving instructions that a computer understands

property: a value that belongs to an object

radius: the distance from the center of a circle to its edge

thumbnail: a very small version of an image

typeface: a family of fonts. A font is a style of letters, numbers, and symbols.

value: data such as numbers or text that can be manipulated by a computer program

variable: a container that can store a value

INDEX

PHOTO ACKNOWLEDGMENTS

The images in this book are used with the permission of: © Antonio_Diaz/iStockphoto, p. 4 (girl); © DaniloAndjus/iStockphoto, pp. 22 (kids), 23 (kids); © izusek/iStockphoto, p. 9; © monkeybusinessimages/iStockphoto, p. 5; © SDI Productions/iStockphoto, p. 30; © SerrNovik/iStockphoto, p. 28 (kids); © Vertigo3d/iStockphoto, p. 15; © Vidcode/Mighty Media, Inc., pp. 4, 6, 7, 8, 10, 11, 12, 13, 14, 16, 17, 18, 19, 20, 21, 22, 23, 24, 25, 26, 27, 28, 29, 29.

Cover Photos: © SerrNovik/iStockphoto; © Vidcode/Mighty Media, Inc.

Design Elements: © Alexey Bezrodny/iStockphoto (arrow); © artishokcs/iStockphoto (pixel background); © filborg/iStockphoto (browser window); © Mighty Media, Inc. (blocks)